Family	Card	Scientific name	Common name
Andrenidae (mining bees) Tiny to medium size, ground nesting; most with 2 vertical facial fovea (depressions with velvety hairs) between eyes; scopae mainly on basal segments of hind leg up to tibia	1	*Andrena* spp.	Mining bees
Apidae Large, diverse family; commonly but not always large bodied; pollen carried in corbiculae or on scopae on hind legs	2	*Anthophora* spp.	Digger bees
	3	*Apis mellifera*	Honey bee
	4	*Bombus* spp.	Bumble bees
	5	*Ceratina* spp.	Small carpenter bees
	6	*Diadasia* spp.	Cactus bees
	7	*Eucera* spp.	Spring long-horned bees
	8	*Eucera* (*Peponapis*) spp.	Squash bees
	9	*Habropoda* spp.	Digger bees
	10	*Melissodes* spp.	Summer long-horned bees
	11	*Svastra* spp.	Summer long-horned bees
	12	*Xeromelecta* spp.	Cuckoo bee
	13	*Xylocopa* spp.	Carpenter bees

Family	Card	Scientific name	Common name
Colletidae (cellophane, or polyester, bees) Tiny to medium size; use cellophane-like secretions to line nests; forked tongue; scopae mainly on hind legs or entirely absent (pollen carried internally)	14	*Colletes* spp.	Polyester bees
	15	*Hylaeus* spp.	Masked bees
Halictidae (sweat bees) Tiny to medium size; distinct arched basal vein in forewing; scopae mainly on hind legs, sometimes extending toward side of thorax	16	*Agapostemon* spp.	Ultra green sweat bees
	17	*Halictus* spp.	Sweat bees
	18	*Lasioglossum* spp.	Sweat bees
Megachilidae (leafcutting bees) Large, diverse family; only 2 submarginal cells on forewing; scopae on underside of abdomen	19	*Anthidium* spp.	Wool carder bees
	20	*Ashmeadiella* spp.	Hairy belly bees
	21	*Coelioxys* spp.	Cuckoo bee
	22	*Hoplitis* spp.	Hairy belly bees
	23	*Megachile* spp.	Leafcutting bees
	24	*Osmia* spp.	Mason bees

Christopher Jadallah in a bee garden established by the Urban Bee Lab at Jim Lloyd-Butler Ranch, Ventura County.

The UC Berkeley Urban Bee Lab

The UC Berkeley Urban Bee Lab has been conducting research on native bee and pollinator-plant interactions for 30 years under the guidance of Dr. Gordon Frankie. We drew the information here from that research and over 20 years of surveying bees in California urban gardens—gardens in diverse climates throughout the state, from Santa Cruz to Bishop, Palm Desert to Ukiah—and our own experimental bee garden on the UC Berkeley campus.

Our surveys yielded thousands of records of hundreds of native bee species visiting California gardens. Some are tiny, others are large and their weight may bend the flower stalk as they forage. Some cut out holes from leaves or collect leaf hairs for their nests. They are different colors and have distinct foraging patterns. We have compiled here some of the most common bees you are likely to see in urban gardens.

How to Use This Guide

The information in this guide is based on more than 20 years of observational surveys on bees in urban environments conducted by the UC Berkeley Urban Bee Lab. The featured bees commonly occur in urban gardens throughout California. This guide can help you identify bees to the genus level (identification to the species level requires a microscope). For information on the flowers they visit, consult the companion guide, *Common Plants to Attract Native Bees* (Chase et al. 2023). Each bee card has these sections:

Names provides the common, genus, and family name; some bees in different genera have the same common name.

General Appearance describes the bee's size, color, shape, and sexual dimorphism (differences between sexes).

Distribution and Richness indicates the number of known species found in California, how many have been recorded visiting urban gardens, and which are widespread and common.

Flight Season notes which months the adult bee is active (variations in weather and climate may slightly alter these times) and flight behavior where relevant.

Nesting provides information on the bee's sociality and nesting behavior.

Pollen Transport describes the part of the body where the bee stores its collected pollen: legs, abdomen, or crop.

Floral Hosts names the bee's preferred flowers for foraging.

Native Bees

There are around 20,000 described species of bees worldwide. The United States and Canada are home to roughly 4,000 species. Nearly 1,600 species, about 40% of those in the United States and Canada, can be found in California. The honey bee (*Apis mellifera*), widely known and one of the most common bee species in California gardens, is not a native bee—it was brought here by European settlers, in 1853.

California's 1,600 species of native bees are diverse in appearance and size, oftentimes looking very different from a typical honey bee. Each species is unique in its nesting and foraging strategy, using vastly different nesting and floral resources to provide shelter and food for its brood.

While most bees are solitary and do not operate within a hive as the social honey bee does, our native bees are still extremely efficient pollinators. Their diversity ensures that California's wide array of flowering plants and crops are pollinated and allows endless research exploring the intricacies of their life histories and ecologies.

This guide focuses on native California bees, but we included the honey bee (*Apis mellifera*) due to its abundance in California gardens. A nonnative wool carder bee (*Anthidium manicatum*) and leafcutting bee (*Megachile rotundata*) are also briefly mentioned, as they are also commonly found in gardens in California.

Solitary Bees, Social Bees

Most California bee species, over 75%, are solitary—a single female bee does all the work of building a nest and provisioning it with pollen and nectar to feed her brood. Some female bees might share a communal nest or situate their nests close together (aggregation), but they rear and provide for only their own offspring. Most of the bees featured in this publication are solitary bees; the social bees featured are the honey bee (*Apis mellifera*), bumble bees (*Bombus* spp.), and some species of sweat bees (e.g., *Halictus* spp.). Look for information on sociality in the "Nesting" section of each bee card.

Social bees make up only about 10% of all bee species in the world. Social bees have 3 characteristics that solitary bees do not: social castes (queens and workers), cooperative brood care (female bees rear offspring that are not their own), and overlapping generations. Some social bees, including honey bees, are eusocial bees—a female worker bee will never become a reproducing, egg-laying female. Other social bees are cooperative breeding bees—the caste system is not permanent; a female bee might begin her life rearing offspring that are not her own and then later lay eggs of her own.

Bee Life Spans

Most of a solitary bee's life is spent in the nest as it transforms from egg to adult. A developing bee goes through 5 larval instars (developmental stages), with its final instar called the prepupa stage. Most bees overwinter in this prepupa stage and pupate in spring into an adult. Some bee groups, including mining bees (*Andrena* spp.), overwinter as adults and emerge earlier, in late winter.

Generally, solitary adult males are active in the landscape for 4 to 6 weeks, while solitary adult females are active for 6 to 8 weeks, though this can vary among groups. Most female bees mate once and produce 1 generation per year, which encourages males to emerge before females (protandry). Once females mate and provision a brood cell, they lay a single egg in the cell. Females then continue to build and provision each cell with resources and a single egg until they close the nest. The total number of eggs laid is dependent on bee species, available floral resources, and other factors. The adult life of solitary bees synchronizes with their preferred floral resources. Many are specialist foragers.

Social bees have an adult life span that can last several months and often are generalist foragers. With multiple, overlapping generations, they are present in the garden most of the year.

Bee Nesting

One can find ground-nesting bees, sometimes called soil-nesting bees, in old beetle burrows, active termite colony mounds, alongside riverbanks, and in sand, clay, sandstone, or rocky outcroppings. The nest is comprised of a main vertical burrow with lateral tunnels that lead to brood cells, where the egg and provisions are and the larva develops when the egg hatches. The arrangement of these cells and length of lateral tunnels and the main burrow vary depending on bee genus and species.

Other bees nest in cavities. There are 2 types of cavity-nesting bees: One type use their mandibles to chew and excavate their nests in wood and inside the pith of stems (like blackberry stems). Depending on bee species, they nest in solid wood or decaying soft wood. The structure of these nests is often linear, with partitions (most commonly of wood debris) between brood cells, and male eggs in the cells close to the entrance because adult males usually emerge first.

The other type of cavity-nesting bees nest in preexisting holes or cavities in wood and inside hollow stems. They may also nest in preexisting ground holes and other already constructed tunnels that another organism excavated.

Cuckoo bees (see *Coelioxys* and *Xeromelecta* spp.) do not build or provision their own nests, but instead lay their eggs in the nests of other bees.

Ground nest

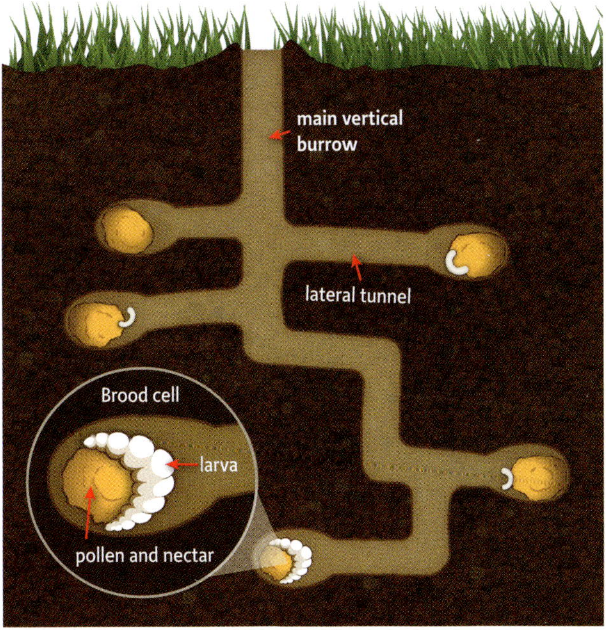

Cavity nest

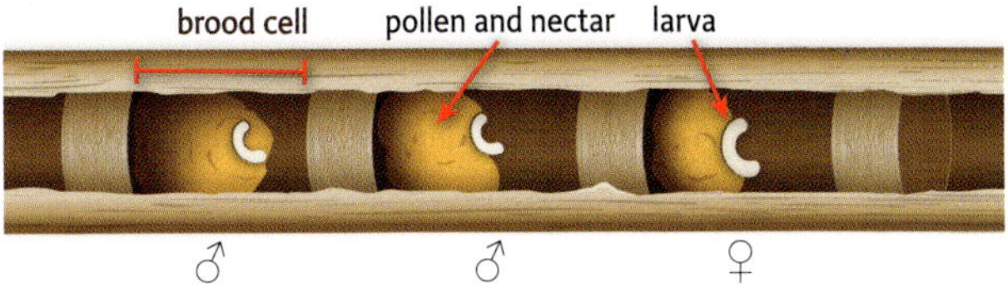

Cuckoo Bees

Cuckoo bees (see *Coelioxys* and *Xeromelecta* spp.) are brood parasites, commonly called cleptoparasites. They make up around 15% of all known bee species worldwide. When a female cuckoo bee is ready to lay an egg, she enters another bee's nest and deposits her egg inside of a cell that has already been provisioned with nectar and pollen. To secure those provisions for her own offspring, the female cuckoo bee will kill the host bee's egg or, once her own egg develops into a larva, it will kill the host egg.

While this strategy seems invulnerable, cuckoo bees are host-specific—they do not enter the nest of any bee species. Not only does a female cuckoo bee need to locate a nest of a specific host species, but the survival of her offspring also depends on the nest having been successfully built and provisioned by the host bee. Additionally, if a cuckoo bee is detected, the host bee may attempt to fight it off or move the nest if it is early in the season.

Cuckoo bees do not collect their own pollen, so they are mostly hairless, but they have features such as a tough exoskeleton, and sometimes projections on the back of the thorax, that allow them to protect themselves from attack. These features often make them appear more wasplike.

Bee Foraging

Bees forage mostly for nectar and pollen. Nectar provides key sugars, amino acids, and proteins. Male bees harvest nectar to gain energy to find a mate; female bees use it for energy and to bind and moisten the pollen they leave for their brood. Pollen is the main protein resource that female bees systematically collect to feed their brood. They eat pollen between foraging trips or at the end of the day to fuel egg laying.

Length and timing of foraging differ among bee species. Larger bees tend to start foraging earlier in the morning, as they can better regulate body temperature during the cold.

Bees are either generalists or specialists in their foraging behavior. Generalists collect pollen and nectar from a wide range of plant families, genera, and species. Specialists depend on a single species or a few plant species, one or a few related genera, or a single plant family.

Pollination occurs accidentally during foraging, as pollen from the bee's body falls onto the flower stigma, or pollen is shaken from the flower stamen onto the stigma. But some bees, including carpenter bees (*Xylocopa* spp.) and bumble bees (*Bombus* spp.), can occasionally be seen engaging in nectar robbing—they make holes in the base of the flower to access the nectaries without interacting with the flower's reproductive parts.

Anatomy of a Bee

A bee's body has 3 parts: head, thorax, abdomen. Antennae that extend from the head aid bees in smell and touch. Two large compound eyes and 3 simple lenses near the top of the head aid their vision. Bees have 6 legs and 2 pairs of wings (4 wings total) extending from the thorax.

Characteristics of a female bee. The female bee has a stinger, for defense. She lays eggs through a hole at the end of the abdomen, at the base of the stinger. Ground-nesting females have plates on their hind legs and at the end of their abdomen that help them dig into and move through soil. Females that nest in preexisting cavities typically are slimmer, which improves their search for a nest location. Females that excavate a cavity for their nest have specialized mandibles to chew through wood.

Most female bees have a set of branched hairs, called a scopa, where they store pollen during flight. There is a scopa on each outer hind leg, or, in some bees, on the underside of the abdomen. Female bumble bees (*Bombus* spp.) store their pollen in a corbicula, an area surrounded by hairs on each hind leg. Female bees without the branched hairs store their pollen internally, in a crop. Noting the presence or absence of scopae or corbiculae can help with bee identification.

Characteristics of a male bee. Generally, males are smaller than females and often have a white or yellow mark on their upper lip (clypeus). Male bees often have modifications that help them find and mate with females, such as very long antennae or special modifications of the legs to clasp the female during mating.

The sex of a bee is most easily assessed by observing how it forages—female bees collect pollen slowly and methodically as they move from flower to flower, while male bees dart between flowers, collecting only nectar. Males may be seen patrolling flower patches or nesting areas for receptive females, and driving off other males.

Bee Size Classes

Bees are typically measured using the intertegular span (IT span), which is the distance between where the wings are attached on the thorax. The size classes here are based on the length of the bee's body; we developed them from published research. Information on each bee's size is provided in the "General Appearance" section of the bee cards.

Large bees are larger than 0.5 (1/2) inch. Examples include many bumble bees (*Bombus* spp.) and carpenter bees (*Xylocopa* spp.).

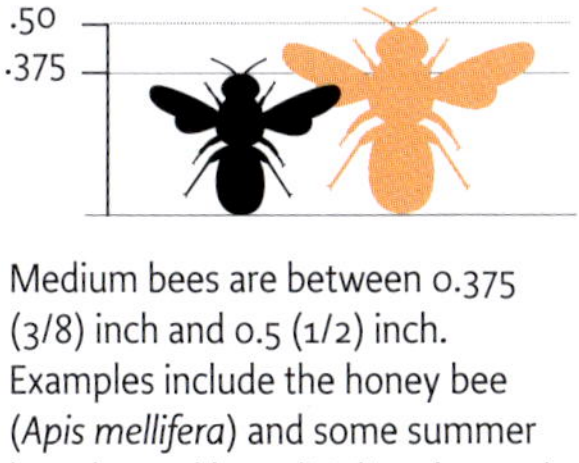

Medium bees are between 0.375 (3/8) inch and 0.5 (1/2) inch. Examples include the honey bee (*Apis mellifera*) and some summer long-horned bees (*Melissodes* spp.).

Small bees are between 0.25 (1/4) inch and 0.375 (3/8) inch. Examples include some sweat bees (*Halictus* spp.) and mining bees (*Andrena* spp.).

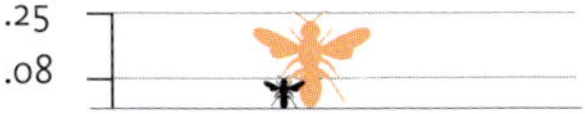

Tiny bees are less than 0.25 (1/4) inch. Examples include small carpenter bees (*Ceratina* spp.) and masked bees (*Hylaeus* spp.).

Glossary

aggregation. Bees nesting near each other in separate nests.

buzz pollination. Vibration of flower anthers by a female bee to release pollen. Only certain larger bees can do this, as it requires a special ability to shiver their flight muscles without moving their wings.

caste. Individuals within a colony that perform the same role (e.g., worker bee).

colony. Queens, workers, and drones living together, in a hive or shared nest.

corbicula (pl. corbiculae). Bare, flattened, or slightly concave area surrounded by hairs on the hind leg of bees into which moist pollen is packed for transport.

corolla. Collective term for flower petals, especially when fused.

crop. The forepart of a bee's digestive system in which some bees store pollen and nectar.

feral bee. Wild honey bee not living in a managed hive.

generalist forager. Bee that collects pollen from many different flowering plant families, genera, and species.

managed pollinator. Pollinator whose nest or hive is transported into an agricultural area to increase crop pollination.

nectar robbing. A behavior in which bees pierce or chew holes into the base of a flower to access the nectaries without interacting with the plant's reproductive parts. Secondary nectar robbers, such as the honey bee (*Apis mellifera*), use the holes made by primary nectar robbers.

race. Informal rank (below subspecies) given to groups of honey bees with shared traits and characteristics; similar to how we think of dog breeds.

scopa (pl. scopae). Set of branched hairs, often densely packed on hind legs, that secure pollen grains for transport.

specialist forager. Bee that restricts pollen collection to 1 or a few closely related species or genera of flowers.

tibia (pl. tibiae). The 4th, or outer, segment of a bee's leg.

Bibliography

Chase, M. H., et al. 2023. Common plants to attract native bees. Oakland: University of California Agriculture and Natural Resources Publication 3557.

Danforth, B., et al. 2019. The solitary bees: Biology, evolution, and conservation. Princeton University Press.

Embry, P. 2018. Our native bees: North America's endangered pollinators and the fight to save them. Portland: Timber Press.

Frankie, G., et al. 2014. California bees & blooms: A guide for gardeners and naturalists. Berkeley: Heyday Books.

Grissell, E. 2010. Bees, wasps, and ants: The indispensable role of Hymenoptera in gardens. Portland: Timber Press.

LeBuhn, G. 2013. Field guide to the common bees of California: Including bees of the Western United States. Berkeley: University of California Press.

The Xerces Society. 2016. 100 plants to feed the bees: Provide a healthy habitat to help pollinators thrive. North Adams, MA: Storey Publishing.

Ullmann, K., et al. 2010. Citizen scientist pollinator monitoring guide. Portland: Xerces Society for Invertebrate Conservation.

Williams, P., et al. 2014. Bumble bees of North America: An identification guide. Princeton: Princeton University Press.

Wilson, J., and O. Messinger Carril. 2015. The bees in your backyard: A guide to North America's bees. Princeton: Princeton University Press.

Online Resources

The Great Sunflower Project, www.greatsunflower.org.

UC Berkeley Urban Bee Lab, www.helpabee.org.

Pollinator Partnership, www.pollinator.org.

Xerces Society for Invertebrate Conservation, www.xerces.org.

1. MINING BEES

Female *Andrena nigrocaerulea* with blue pollen on hind leg scopae and yellow pollen on thorax.

Female *Andrena angustitarsata* visiting a *Ceanothus* sp. blossom.

Female *Andrena* sp. collecting pollen from a *Calochortus* sp. flower.

Female *Andrena subtilis* with characteristic pollen-covered hind leg.

1. Mining bees

Andrena spp., Andrenidae family

General Appearance: Small to large, long, and slender, often with pale hair on the thorax and pale stripes on the abdomen. The abdomen or entire body may have a metallic sheen. Females have depressions covered with velvety hairs (facial foveae) that run vertically between the eyes, though shape and size of foveae vary between species.

Distribution and Richness: 225 species in California, 35 of which have been recorded visiting urban gardens. 5 species are widespread: *Andrena angustitarsata, A. auricoma, A. candida, A. knuthiana,* and *A. melanochroa.*

Flight Season: February into June, with peak activity from March into May.

Nesting: Females are solitary and nest in soil, usually in flat, bare ground. Nest entrances are often surrounded by concentric mounds of excavated soil.

Pollen Transport: In scopae on the outer part of the hind legs, augmented by long, curled hairs near the base of the legs and corbicular areas at the back of the thorax. Branched hairs of scopae appear closer to the base of the hind leg than on other bees (e.g., *Halictus* spp.).

Floral Hosts: Most are generalist foragers of pollen, often visiting *Ceanothus* and *Phacelia* and occasionally *Encelia, Gaillardia,* and *Salvia* species.

2. DIGGER BEES

Sleeping males of *Anthophora californica.*

Female *Anthophora urbana* foraging on *Phacelia campanularia.*

Male *Anthophora urbana* foraging on *Melilotus officinalis.*

Female *Anthophora pacifica* on *Arctostaphylos* sp.

2. Digger Bees

Anthophora spp., Apidae family

General Appearance: Large to small, stout bodied, gray with black stripes on the abdomen. Females are larger and have a dark face, while males are smaller and more elongate with pale markings on the lower part of the face.

Distribution and Richness: 40 species in California, 6 of which have been recorded visiting urban gardens. *Anthophora urbana* is the most widespread, but *A. californica* and *A. curta* also occur in urban gardens.

Flight Season: February to June, with peak activity in March and April for most species. *Anthophora urbana* is an exception; it usually flies from May into October, with peak flight activity from June into September. Often appear to "hop" from flower to flower while foraging.

Nesting: These bees are solitary and nest in the soil. Nests are often located in flat, bare ground and sometimes in vertical banks along the sides of roads or trails.

Pollen Transport: In scopae on the hind legs.

Floral Hosts: Females are generalist foragers and commonly take nectar from plants in the Lamiaceae family, such as *Lavandula, Nepeta*, and *Salvia* species. *Anthophora urbana* commonly visits *Cosmos, Echium, Perovskia, Phacelia, Teucrium*, and *Vitex* species.

3. HONEY BEE

Apis mellifera worker visiting *Gilia capitata*. Note the blue pollen in the bee's corbicula.

Apis mellifera worker on *Aster* x *frikartii*. Note the corbicula: flattened hind tibia surrounded by hairs.

Apis mellifera are important agricultural pollinators. Here one visits a

An *Apis mellifera* worker bee drinking water.

3. Honey Bee

Apis mellifera, Apidae family

General Appearance: Worker honey bees are sterile females, medium size, thick bodied, and yellowish to dark brown with striped abdomens. Males (drones) are larger than workers, with enlarged eyes that meet at the top of the head. Queens are more elongate than workers and lack pollen transport structures. Workers are the only caste found foraging at flowers.

Distribution and Richness: Native to Europe, introduced to continents across the globe. Colonies are either feral or managed by beekeepers. The Africanized race (*Apis mellifera scutellata*) has become common throughout most of Southern California and is difficult to distinguish from other honey bees, except by noting increased defensive behavior displayed near nests. When a colony of Africanized honey bees is disturbed, they will defend their nests over a greater area than European honey bees.

Flight Season: Year round.

Nesting: Honey bees are highly social and nest in large cavities in trees, soil, buildings, or managed hive boxes. Occasionally, they construct exposed nests from tree limbs.

Pollen Transport: As moistened pellets in corbiculae on the hind tibiae.

Floral Hosts: Workers are broad generalist foragers—they visit a large variety of native and nonnative flowers.

4. BUMBLE BEES

Bombus vosnesenskii workers at their underground nest entrance.

Visible pollen loads in corbiculae of a *Bombus californicus* worker.

Bombus melanopygus female visiting *Ribes sanguineum* flowers.

Bombus vosnesenskii worker foraging.

4. Bumble Bees

Bombus spp., Apidae family

General Appearance: Medium size to very large and stout bodied, often with copious predominantly black hair and a few sharply contrasting yellow, red, and/or white hair bands.

Distribution and Richness: 27 species in California. 3 are common: *Bombus vosnesenskii, B. melanopygus,* and *B. californicus.*

Flight Season: Present most of the flowering season. Queens of some species emerge from hibernation as early as late December to start nests and begin producing workers. Workers are common from spring into fall. New queens and males are produced at the end of the colony season.

Nesting: Bumble bees are social. They typically nest underground in abandoned rodent burrows or in tufts of grass. Some, like *Bombus melanopygus*, may nest in abandoned bird nests or birdhouses.

Pollen Transport: As moistened pellets in corbiculae.

Floral Hosts: Female bumble bees are broad generalist foragers for pollen. Different species have different tongue lengths, which affects their choice of nectar sources. Short-tongued species may chew holes in the base of flowers to access the nectar (nectar robbing). *Arctostaphylos, Ceanothus, Eschscholzia, Lavandula, Nepeta, Salvia,* and *Vitex* species are attractive floral hosts. Bumble bees can buzz-pollinate flowers, and may pollinate tomato plants this way.

5. SMALL CARPENTER BEES

Male *Ceratina* sp. showing ivory-colored mark on the face.

Female *Ceratina* sp. displaying characteristic tapered abdomen that aids in identification.

Female *Ceratina acantha* visiting flower of *Diplacus* (*Mimulus*)

Female *Ceratina* sp. with pollen from a wild rose stored on leg scopa.

5. Small Carpenter Bees

Ceratina spp., Apidae family

General Appearance: Tiny to small, rather hairless, shiny, and dark metallic blue green. Abdomen is a distinctive shield shape, with a pointed tip.

Distribution and Richness: 13 species in California, 5 of which have been recorded visiting urban gardens. 2 are widespread and common, *Ceratina acantha* and *C. nanula.*

Flight Season: April through August, although they can sometimes be active earlier or later in year. They tend to be quick and deliberate fliers.

Nesting: Females are solitary and construct a nest by excavating the center of a pithy or soft-core dead stem of a plant such as *Sambucus, Rubus, Brassica, Helianthus, Perovskia, Phlomis,* or *Salvia mellifera.*

Pollen Transport: In sparse scopae on the outer part of the hind legs.

Floral Hosts: Females are generalist foragers, often found on plants in the Asteraceae family, such as *Aster, Erigeron,* and *Grindelia* species. They visit flowers of plants in the Lamiaceae family, especially *Nepeta* and *Perovskia* species, for nectar.

6. CACTUS BEES

Female *Diadasia* sp. on flowers of *Sphaeralcea ambigua*.

Female *Diadasia ochracea*, with pollen-covered legs, on *Malacothamnus* sp.

Female *Diadasia ochracea* foraging on *Malacothamnus* sp.

Characteristic turrets define the nest entrances of *Diadasia enavata*.

6. Cactus Bees

Diadasia spp., Apidae family

General Appearance: Small to large, stout bodied, light brown with striped abdomen. The head is narrower than the thorax, and the top of the head is round. Females are larger than males. The tip of the female's abdomen is upturned, while that of the male is downturned. Some species are densely covered with velvety hairs.

Distribution and Richness: 20 species in California, 8 of which have been recorded visiting urban gardens. *Diadasia bituberculata* is widespread.

Flight Season: April to August.

Nesting: Females are solitary and nest in flat, bare soil. Nest entrances are often adorned with characteristic turrets formed from cemented soil particles. Aggregated nesting, where bees construct individual nests in clusters, is common.

Pollen Transport: In scopae on the hind legs.

Floral Hosts: In most species, the females specialize collecting pollen from plants of the Malvaceae family, such as *Sphaeralcea* species. A few specialize their pollen foraging on *Opuntia, Calystegia, Convolvulus,* or *Helianthus. Diadasia* are less selective about nectar, accepting many different sources.

7. SPRING LONG-HORNED BEES

Male *Eucera lunata*, displaying the characteristic yellow face and long black antennae, drinking nectar from *Arctostaphylos* sp.

Female *Eucera frater albopilosa* foraging on *Nepeta* sp.

Male *Eucera frater albopilosa*, also with characteristic yellow face and

Female *Eucera actuosa* resting.

7. Spring Long-Horned Bees

Eucera spp., Apidae family

General Appearance: Medium to large, stout bodied, with gray hair on the thorax. Some have uniform pale stripes on the abdomen. Females are larger, with a dark face; males are smaller and slender, with yellow markings on the face and very long, uniformly dark antennae.

Distribution and Richness: 25 species in California, 6 of which have been recorded visiting urban gardens. 2 of these are squash bees formerly classified in the genus *Peponapis.*

Flight Season: March to June.

Nesting: Females are solitary and nest in the soil.

Pollen Transport: In scopae on the outer part of the hind legs.

Floral Hosts: Females are often generalist foragers for pollen. Some preferred plants include *Erysimum, Nepeta,* and *Lavandula* species.

8. SQUASH BEES

Male *Eucera pruinosa* resting inside a squash flower.

Male *Eucera pruinosa* often lie in wait for females inside squash flowers in the early morning.

Female *Eucera pruinosa* inside squash flower.

Note the longer antennae and pale mark on the face of this male *Eucera*

8. Squash Bees

Eucera (Peponapis) spp., Apidae family

General Appearance: Medium size, stout bodied, with brownish-orange hair on the thorax and brownish stripes on the abdomen. They look similar to **honey bees**.

Distribution and Richness: There are 2 species of squash bees in California, only 1 of which, *Eucera pruinosa* (*Peponapis pruinosa*), has been recorded visiting urban gardens. *Eucera pruinosa* is fairly common throughout California wherever squash and gourds are planted. The other species, *E. timberlakei* (*P. timberlakei*) is found only in southeastern California and is not abundant.

Flight Season: June to September, with peak flight from June to August.

Nesting: Females are solitary and create their nest in flat, bare ground.

Pollen Transport: In pale brownish scopae on the outer part of the hind legs.

Floral Hosts: Females gather pollen from flowers of the genus *Cucurbita*, a group that includes many food plants such as squash and pumpkins. They visit other plants for nectar, including vine crops such as melons and cucumbers, and *Ipomoea* species. Males sleep in closed *Cucurbita* species flowers from midmorning to just before dawn the following day.

9. DIGGER BEES

Female *Habropoda depressa* flying toward flowers of *Ceanothus* sp.

Female *Habropoda miserabilis* foraging on *Phacelia* sp.

Mating pair of *Habropoda depressa*. Note the male's distinguishable white

Female *Habropoda depressa* leaving her nest.

9. Digger Bees,

Habropoda spp., Apidae family

General Appearance: Large, stout bodied, gray, and hairy. Females are larger than males and have a dark face. Males have pale markings on the lower part of the face and often at the base of the antennae and mandibles.

Distribution and Richness: 12 species in California, 2 of which have been recorded visiting urban gardens. *Habropoda depressa* is widespread and common.

Flight Season: February into May, with peak activity in March and April. They are fast and noisy fliers.

Nesting: Females are solitary and nest in soil, often in vertical banks but also in flat ground. Aggregated nesting is common.

Pollen Transport: In white scopae on the outer part of the hind legs.

Floral Hosts: Females are often found on early-blooming plants such as *Arctostaphylos* and *Ceanothus* species, but they also gather pollen from plants that are not typically pollinated by bees, such as oak trees. They are among the few bee groups that can buzz-pollinate flowers.

10. SUMMER LONG-HORNED BEES

Males of *Melissodes robustior* sleeping in a cluster on a *Cosmos* flower.

Female *Melissodes robustior* displaying pollen stored in scopa on hind tibia.

Male *Melissodes robustior* displaying characteristic long antennae.

Female *Melissodes robustior* foraging on a *Gaillardia* sp.

10. Summer Long-Horned Bees

Melissodes spp., Apidae family

General Appearance: Medium to large, stout bodied, usually with gray hair on the thorax and pale hair bands on the abdomen. Males usually have yellow markings on the face and have very long, bicolored antennae.

Distribution and Richness: 50 species in California, 9 of which have been recorded visiting urban gardens. *Melissodes robustior, M. tepida timberlakei,* and *M. lupina* are widespread and common.

Flight Season: May to September, with peak flight in late June to early August.

Nesting: Females prefer flat, bare ground for their solitary nest, though they sometimes nest in aggregations.

Pollen Transport: In scopae on the hind legs. Pollen loads are often copious and brightly colored, a very distinguishable feature of these bees.

Floral Hosts: Females gather pollen from plants in the Asteraceae family, such as *Aster, Bidens, Coreopsis, Cosmos, Encelia, Gaillardia, Helianthus,* and *Rudbeckia* species. Flowers of *Scabiosa* and species in the Lamiaceae family, such as *Lavandula* and *Nepeta* species, provide nectar. Males can often be found sleeping in clusters on floral heads in the early morning or late evening.

11. SUMMER LONG-HORNED BEES

Male *Svastra obliqua expurgata* nectaring on *Rudbeckia* sp.

Female *Svastra obliqua expurgata* foraging.

Female *Svastra obliqua expurgata* foraging.

Male *Svastra obliqua expurgata* nectaring on *Echinacea purpurea*.

11. Summer Long-Horned Bees

Svastra spp., Apidae family

General Appearance: Medium to large, stout bodied, usually with gray hair on the thorax and pale, irregular stripes on the abdomen. Once considered to be within the genus *Melissodes*, *Svastra* were separated into their own genus, due to subtle characteristics distinguishable only by microscope.

Distribution and Richness: 7 species in California. Only *Svastra obliqua expurgata* has been recorded visiting urban gardens.

Flight Season: May to September, with peak flight in late June to early August.

Nesting: Females are solitary but nest in aggregations in flat, bare ground. They sometimes make use of cracks in the soil to penetrate heavier clay soils.

Pollen Transport: In scopae on the hind legs.

Floral Hosts: When foraging for pollen, females strongly prefer plants in the Asteraceae family, especially *Helianthus*, *Gaillardia*, and *Ratibida* species.

12. CUCKOO BEE

Male *Xeromelecta californica* foraging for nectar from *Aster* x *frikartii* 'Monch'.

Male *Xeromelecta californica* from another angle.

Female *Xeromelecta californica* at nest entrance of its host, *Anthophora bomboides stanfordiana*.

Female *Xeromelecta californica* resting on a stem.

12. Cuckoo Bee

Xeromelecta spp., Apidae family

General Appearance: Medium size, with grayish-white hairs on the head and thorax. Distinct white stripes cover the tapered, mostly hairless abdomen.

Distribution and Richness: 2 species in California, with only *Xeromelecta californica* recorded visiting urban gardens. *Xeromelecta californica* is widespread but not abundant. The other species, *X. larreae*, occurs in Southern California but is neither widespread nor abundant.

Flight Season: Emerge in conjunction with host bees, often July into October.

Nesting: As cleptoparasites, these cuckoo bees do not construct their own nests. Rather, they lay their eggs in the nests of other bees, especially digger bees (*Anthophora* spp).

Pollen Transport: These bees do not collect pollen for their brood since they lay their eggs in already-provisioned nests.

Floral Hosts: Forage for nectar only, often seen in plants of the Asteraceae and Lamiaceae families, such as *Cosmos, Rudbeckia, Perovskia*, and *Lavandula* species.

13. CARPENTER BEES

Male *Xylocopa sonorina* with its characteristic golden color.

Female *Xylocopa tabaniformis orpifex* forages for nectar from *Salvia.*

Male *Xylocopa tabanaiformis orpifex* visiting flowers of *Cercis occidentalis.*

Female *Xylocopa sonorina* "nectar robbing" from a flower.

13. Carpenter Bees

Xylocopa spp., Apidae family

General Appearance: Large, shiny, and stout bodied, with sparse hair. Females are mostly black. Males often have some yellow hairs, especially on the front of the thorax, and yellow coloring on the lower part of the face. Males of *Xylocopa sonorina* (*X. varipuncta*) are completely golden brown, with greenish eyes.

Distribution and Richness: 3 species in California, all of which have been recorded visiting urban gardens: *Xylocopa californica*, *X. tabaniformis orpifex*, and *X. sonorina*. All are widespread and common.

Flight Season: March to October, with peak activity from April into July.

Nesting: Solitary and long-lived, they often overwinter with other adults in their natal nest. Females create new nests with their strong mandibles by chewing holes in soft or decaying wood, especially in untreated pine or redwood structures.

Pollen Transport: In scopae on hind legs.

Floral Hosts: Females are generalist foragers. In urban gardens, they are often found visiting plants in the Fabacaeae family, such as *Cercis* species, *Parkinsonia* species, and *Wisteria sinensis*, for pollen and nectar and plants in the Lamiaceae family, such as *Lavandula* and *Salvia* species, for nectar. They often rob nectar by chewing holes into the base of flowers and can buzz-pollinate.

14. POLYESTER BEES

Male *Colletes* sp. displaying long antennae, white face hairs, and slender hind legs.

Female *Colletes* sp. foraging on *Eriogonum* sp.

Female *Colletes hyalinus gaudialis* foraging on *Erigeron glaucus*.

Female *Colletes fulgidus longiplumosus* covered in pollen while foraging on *Eriogonum* sp.

14. Polyester Bees

Colletes spp., Colletidae family

General Appearance: Small to medium, long, slender, gray or brown, with a triangular-shaped head and hair bands on the abdomen. In females, the abdomen is tapered to a point. Males are more slender than females, with dense white facial hair.

Distribution and Richness: 40 species in California, 4 of which have been recorded visiting urban gardens. *Colletes hyalinus gaudialis* is fairly widespread.

Flight Season: March into May.

Nesting: Females are solitary and nest in soil. Some make aggregated nests in vertical sandstone banks. Sometimes also called membrane bees, female *Colletes* secrete a polyester-like material from a gland in the abdomen (Dufour's gland), which they use to line nest tunnels.

Pollen Transport: In scopae, which extend from the thorax to the tip of the hind legs.

Floral Hosts: Most species females are generalist foragers of pollen, but some show a preference for plants in the Asteraceae family, such as *Solidago* species.

15. MASKED BEES

Male *Hylaeus rudbeckiae* resting on a blossom.

Female *Hylaeus mesillae* foraging on flowers.

Male *Hylaeus punctatus* with 1 solid yellow mark in the middle of the face.

Female *Hylaeus punctatus* with lateral yellow marks on each side of the face.

15. Masked Bees

Hylaeus spp., Colletidae family

General Appearance: Tiny, mostly hairless, and black. Females usually have a pair of vertical yellow marks on the face, while males have a broader yellow mark in the center of the face, in addition to the lateral face stripes.

Distribution and Richness: 26 species in California, 11 of which have been recorded visiting urban gardens. *Hylaeus mesillae, H. polifolii, H. punctatus,* and *H. rudbeckiae* are widespread and common.

Flight Season: March into October, with peak activity from July into September.

Nesting: These bees are solitary; they construct a nest in a preexisting tunnel such as a beetle burrow or hollow stem.

Pollen Transport: Internally in the crop, the forepart of the digestive system, so there is no visible external pollen-holding structure.

Floral Hosts: Females are broad generalist foragers and can be found visiting *Achillea, Aster, Ceanothus, Eriogonum, Phacelia, Salvia,* and *Solidago* species.

16. ULTRA GREEN SWEAT BEES

Male *Agapostemon texanus* cleaning its antennae while resting on a *Cosmos* sp. flower.

Female *Agapostemon texanus* visiting *Erigeron glaucus*.

The striped abdomen of male *Agapostemon texanus* individuals helps differentiate them from females.

Female *Agapostemon texanus* emerging from her underground nest.

16. Ultra Green Sweat Bees

Agapostemon spp., Halictidae family

General Appearance: Medium to small, slender bodied, and bright metallic green. Females are usually entirely metallic green, though they sometimes have a brown or black abdomen. The base of mandibles is usually pale yellow. Males have a green head and thorax but a yellow-and-dark striped abdomen. Their mandibles are usually entirely pale yellow.

Distribution and Richness: 5 species in California, 2 of which have been recorded visiting urban gardens. *Agapostemon texanus* is widespread and common.

Flight Season: March into October, with peak flight activity from May into September. Some have multiple generations per year.

Nesting: Females are solitary and nest in the soil, most often in bare, flat ground.

Pollen Transport: In scopae covering most of the hind legs, from the tip to the base.

Floral Hosts: Males and females are broad generalist foragers. Females are often found visiting flowers in the Asteraceae family, and both sexes visit *Cosmos, Erigeron, Perovskia, Rosmarinus,* and *Rudbeckia* species.

17. SWEAT BEES

Female *Halictus tripartitus* resting on a blossom of *Eriogonum grande* var. *rubescens.*

Female *Halictus ligatus* foraging on a blossom.

Female *Halictus farinosus* collecting pollen from *Eschscholzia californica.*

Female *Halictus tripartitus* near her nest entrance in the ground.

17. Sweat Bees

Halictus spp., Halictidae family

General Appearance: Small to medium, long, and dark, with pale hair bands at the end of each abdominal segment.

Distribution and Richness: 6 species in California, 4 of which have been recorded visiting urban gardens. *Halictus farinosus, H. ligatus*, and *H. tripartitus* are widespread and common.

Flight Season: Females appear from March into October; males do not appear until June. Peak activity is from June into September.

Nesting: Species span a spectrum of sociality; some are solitary, and others are highly social and form colonies. They tend to nest in the ground.

Pollen Transport: In scopae extending from the sides of the thorax to the tip of the outer hind legs.

Floral Hosts: They are broad generalist foragers when foraging for pollen. Quite common on plants in the Asteraceae family, they especially prefer *Aster, Bidens, Coreopsis, Erigeron, Helianthus*, and *Solidago* species.

18. SWEAT BEES

Small female *Lasioglossum* (*Dialictus*) sp. collecting pollen from anthers of *Phacelia campanularia*.

Female *Lasioglossum* (*Dialictus*) sp. carrying pollen from *Fremontodendron californica* in hind leg scopa.

Female *Lasioglossum kinkaidii* on flower of *Grindelia* sp.

Female *Lasioglossum* (*Dialictus*) sp. visiting a tiny weed flower.

18. Sweat Bees

Lasioglossum spp., Halictidae family

General Appearance: Tiny to medium, long, slender, dark in color. Females have a dark face and are larger than males. Males are more slender than females and have yellow facial markings.

Distribution and Richness: There are 3 major subgenera: *Lasioglossum* has 13 species in California, with 9 recorded in urban gardens. *Dialictus* has nearly 40 species in California, with 2 named species and 10 morphospecies (recognized but unnamed) recorded in urban gardens. *Evylaeus* has 30 species in California, also with 2 named and 10 morphospecies recorded in urban gardens.

Flight Season: Bees of the subgenus *Dialictus* display social polymorphism—they can be weakly eusocial or solitary depending on climate. They can be observed from March through September. Most in the subgenera *Evylaeus* and *Lasioglossum* are solitary and fly from March into August.

Nesting: Females nest in the soil, preferring flat, bare ground. Nests are often aggregated.

Pollen Transport: In scopae extending from the tip of outer hind legs in toward the sides of the thorax. Occasionally, females have additional branched hairs beneath the abdomen.

Floral Hosts: Females are broad generalist foragers, although some exhibit preference for plants in the Asteraceae family; often observed on *Erigeron* and *Eschscholzia* species.

19. WOOL CARDER BEES

Female *Anthidium manicatum* collecting plant fibers to line her nest.

Male *Anthidium manicatum* nectaring on *Stachys byzantina*.

Female *Anthidium manicatum* foraging on *Salvia chamaedroides*.

Mating pair of *Anthidium maculosum*. Males are territorial and can be seen patrolling flower patches for females.

19. Wool Carder Bees

Anthidium spp., Megachilidae family

General Appearance: Small to large, stout bodied, and black with yellow markings. Males have prominent tooth-like projections at the tip of the abdomen.

Distribution and Richness: 22 species in California, 8 of which have been recorded visiting urban gardens. *Anthidium manicatum* (nonnative) and *A. maculosum* (native) are widespread and common.

Flight Season: May into October, with peak flight from June into August for most species. Males are very territorial and patrol specific patches of flowers, defending them not only against males of their own species but also against many other insects.

Nesting: Most are cavity nesters; some may excavate their nests in sand. Females are solitary and scrape hairs from the surfaces of leaves and stems to place them in nesting cavities to form a cottonlike filling.

Pollen Transport: On the underside of the abdomen.

Floral Hosts: *Anthidium maculosum* is a generalist forager but most often found visiting *Salvia* species in urban gardens. *Anthidium manicatum* prefers *Calamintha, Linaria,* and *Stachys* species.

20. HAIRY BELLY BEES

Ashmeadiella bucconis denticula foraging on a blossom.

Ashmeadiella rhodognatha resting on the ground.

Male *Ashmeadiella* sp. resting on a cactus spine.

20. Hairy Belly Bees

Ashmeadiella spp., Megachilidae family

General Appearance: Small and chunky, usually black with pale hair bands on the abdomen. Shiny, smooth area at the front sides of the thorax is separated from the rest of the thorax by a small sharp ridge. A few species are reddish.

Distribution and Richness: 42 species in California, 4 of which have been recorded visiting urban gardens. *Ashmeadiella aridula astragali* is relatively widespread.

Flight Season: May into September, with peak season in July and August.

Nesting: Most species are solitary, and nest in a preexisting cavity, using a wide array of nesting materials. To partition brood chambers, they use chewed-up leaf pulp and plant resins.

Pollen Transport: In scopae on the underside of the abdomen.

Floral Hosts: Several are specialist foragers, specializing on *Cryptantha, Penstemon,* or *Phacelia* species. Most encountered in urban gardens are generalist foragers, visiting *Encelia, Erigeron, Linaria, Phacelia, Perovskia,* and *Rosmarinus* species.

21. CUCKOO BEE

Female *Coelioxys rufitarsis* foraging for nectar on *Gaillardia* sp.

Female *Coelioxys rufitarsis* foraging for nectar on *Jacobaea maritima*. Note the exceptionally tapered abdomen.

Male of *Coelioxys rufitarsis* foraging for nectar on a blossom. Note the spines on the tip of the abdomen

Female *Coelioxys rufitarsis* sleeping.

21. Cuckoo Bee

Coelioxys spp., Megachilidae family

General Appearance: Medium size, with gray and black stripes on the abdomen. Females have a tapered, cone-shaped abdomen with a distinctive point at the end; the abdomen of males tapers and ends with a row of teeth, or spines. Mostly hairless.

Distribution and Richness: 16 species in California, 7 of which have been recorded visiting urban gardens. *Coelioxys* species are widespread but not abundant.

Flight Season: Emerge in conjunction with host bees, usually May into October.

Nesting: As cleptoparasites, they do not construct their own nests. Rather, they lay their eggs in the nests of other bees, especially leafcutting bees (*Megachile* spp.).

Pollen Transport: These bees do not collect pollen for their brood since they lay their eggs in already-provisioned nests.

Floral Hosts: Several are broad generalist foragers for nectar, although they exhibit a preference for plants in the Asteraceae family, especially *Aster* x *frikartii, Calendula, Grindelia,* and *Cosmos* species.

22. HAIRY BELLY BEES

Hoplitis coeplandica albomarginata male that was removed from a cavity nest.

Male *Hoplitis albifrona maura* resting.

Hoplitis albifrons maura emerging from a nail hole, where they spent the night.

Male *Hoplitis fulgida platyura* that was found emerging from a nail hole.

22. Hairy Belly Bees

Hoplitis spp., Megachilidae family

General Appearance: Range widely in size and color. *Hoplitis fulgida* is slender and green, while other species are black or black and red. Some have pale hairs on the thorax and hair bands on the abdomen. Some *Hoplitis* species males have unusual and distinct-looking antennae, which are often swollen at the base, thick in the middle, and pointed at the tip.

Distribution and Richness: 70 species in California, 5 of which have been recorded visiting urban gardens. None of these species are widespread or common.

Flight Season: May into September, with peak season in June and July.

Nesting: Most are solitary and nest in a cavity, using chewed-up leaf pulp and plant resins to partition brood chambers.

Pollen Transport: In scopae on the underside of the abdomen.

Floral Hosts: Several are specialist foragers and specialize on *Cryptantha*, *Penstemon*, or *Phacelia* species. Most encountered in urban gardens are generalist foragers, visiting *Encelia*, *Erigeron*, *Linaria*, *Phacelia*, *Perovskia*, and *Salvia* species.

23. LEAFCUTTING BEES

Female *Megachile perihirta* collecting pollen from *Erigeron* sp. Note how it is stored under the abdomen.

Male *Megachile montivaga* nectaring on *Erigeron* sp.

Female *Megachile centuncularis* using petals to close her nest in an artificial nest block.

Female *Megachile rotundata* (nonnative) foraging on *Erigeron karvinskianus*.

23. Leafcutting Bees

Megachile spp., Megachilidae family

General Appearance: Large to small, stout bodied, usually with pale hair on the thorax and stripes of white hairs on the abdomen. The female usually has a triangular abdomen with a pointed tip, and the male's face is covered with dense, pale hairs.

Distribution and Richness: 75 species in California, 23 of which have been recorded visiting urban gardens. 6 of these are widespread.

Flight Season: May into September, with peak activity from June to August.

Nesting: Females are solitary and construct their nest in a tubular preexisting cavity, such as a hollow stem, tree hole, or abandoned beetle burrow in wood. Many use a hole drilled into wood, or a straw, or other manufactured tunnel. They cut pieces from leaves and flower petals to construct their brood cells.

Pollen Transport: In dense scopae on the underside of the abdomen.

Floral Hosts: Females are generalist foragers; however, some species, including *Megachile fidelis*, exhibit a preference for plants in the Asteraceae family, such as *Aster, Cosmos, Erigeron, Helenium,* and *Rudbeckia* species. Other species, including *M. angelarum*, visit mostly *Lavandula, Perovskia,* and *Vitex* species.

24. MASON BEES

Female *Osmia* sp. at her nest entrance inside a hollow stem.

Female *Osmia lignaria* foraging on an apple blossom.

Female *Osmia aglaia* at the entrance of a flower of *Penstemon heterophylla*.

Female *Osmia coloradensis* foraging on *Erigeron glaucus*.

24. Mason Bees

Osmia spp., Megachilidae family

General Appearance: Small to large, stout bodied, with a round head, a larger round thorax, and even larger rounded abdomen. Usually dark metallic blue, but a few are bright metallic green.

Distribution and Richness: 95 species in California, 21 of which have been recorded visiting urban gardens. *Osmia coloradensis*, *O. lignaria propinqua*, and *O. aglaia* are widespread and common.

Flight Season: March into June, except for *O. coloradensis*, which flies into July.

Nesting: Females are solitary and construct a nest in a preexisting cavity, such as a hole drilled in a board, a hollow reed, a hole in cardboard, or a straw. To build the nest, they use mud or chewed-up leaf pulp.

Pollen Transport: In dense scopae on the underside of the abdomen.

Floral Hosts: Females are mostly generalist foragers when foraging for pollen, but some species are specialists. *Osmia coloradensis* prefers plants in the Asteraceae family, such as *Bidens*, *Calendula*, *Coreopsis*, *Encelia*, and *Grindelia* species. *O. aglaia* is often found visiting *Lavandula*, *Penstemon*, and *Phacelia* species. *O. lignaria* often visits *Cercis*, *Berberis*, *Echium*, *Phacelia*, and *Salvia* species and is an important managed pollinator for fruit crops such as apples and cherries.

Acknowledgments: The authors of the 2017 edition were Christopher Jadallah, Gordon Frankie, Robbin Thorp, Rollin Coville, Sara Leon-Guerrero, and Ingrid Feng. Some of the material in that edition was reprinted from *California Bees and Blooms* with the kind permission of Heyday.

To order or obtain UC ANR publications and other products, visit the UC ANR online catalog at https://anrcatalog.ucanr.edu/ or phone 1-800-994-8849. Direct inquiries to

UC Agriculture and Natural Resources
Publishing
2801 Second Street
Davis, CA 95618

Telephone 1-800-994-8849
E-mail: anrcatalog@ucanr.edu

Publication 3552-2 (second edition, 2023)

ISBN-13: 978-1-62711-146-1

LCCN: 2021948017

Design by Sandra Osterman. All bee photographs by Rollin Coville.

UC PEER REVIEWED This publication has been anonymously peer reviewed for technical accuracy by University of California scientists and other qualified professionals. This review process was managed by UC ANR Associate Editor for Pest Management—Urban, Dr. Andrew Mason Sutherland.

Printed in U.S.A.

2.5m-pr-01/23-HW/SO